Milk

Claire Llewellyn

W

FRANKLIN WATTS

LONDON • SYDNEY

First published in 2004
by Franklin Watts
96 Leonard Street
London EC2A 4XD

Franklin Watts Australia
45-51 Huntley Street
Alexandria NSW 2015

Series advisor: Gill Matthews, non-fiction literacy specialist and Inset trainer
Editor: Caryn Jenner
Series design: Peter Scoulding
Designer: James Marks
Photography: Ray Moller unless otherwise credited
Acknowledgements: Nigel Cattlin/Holt Studios: 11, 14, 15, 16, 17, 23bl. Bennett Dean/Eye Ubiquitous:
8-9, 22br. Nigel Dickinson/Still Pictures: 13. Wayne Hutchinson/Holt Studios: 12.
Thanks to our models: Khailam Palmer Mutlu, Jakob Hawker, Reanne Birch, Casey Liu, Thomas Ong.

A CIP catalogue record for this book is available from the British Library

ISBN: 0 7496 5414 7

Printed in Malaysia

Contents

Drinking milk

Most of us drink milk every day. We have it on its own...

in hot drinks...

When do you drink milk?

and on breakfast cereals.

5

Milk is good for us

Milk helps our bodies in many different ways.

▶ *Milk helps us to grow and stay healthy.*

▶ *Milk also gives us energy to work and play.*

Milk is good for our bones and teeth. It makes them very strong.

Cow's milk

Most of the milk we drink comes from cows. Cows live on farms.

▶ Cows usually feed on grass in the fields.

Some cows are kept for their milk. Others are kept for their meat.

Milk for the calf

A cow makes milk when it has a calf. We take some of this milk for ourselves.

Many animals make milk for their babies. Animals that do this are called mammals. Humans are mammals.

▶ *The calf sucks milk from the cow's udder.*

Milking the cows

Farmers milk their cows twice a day. They milk them in the milking parlour.

▲ *Many cows are milked at the same time.*

Cows are not the only animals that give us milk. Goats and sheep do, too.

◀ *A milking machine sucks milk from the udder, just like a calf.*

From the farm to the dairy

The farmer stores the milk in a clean, cold tank. Each morning a tanker takes the milk to a dairy.

▲ *The cows' milk is stored in this tank.*

14

▲ *A tanker collects milk from the farm.*

The farm tank and the tanker keep the milk cold and fresh. Where do you keep your milk at home so it stays cold and fresh?

At the dairy

There are germs in milk. At the dairy, the germs are killed so that the milk is safe to drink.

▶ *These machines heat the milk and kill the germs.*

Germs are tiny living things that can make us ill. What can we do to stop germs spreading?

The milk is put into containers.

Dairy foods

Milk is used to make many different dairy foods.

Butter ▶

▼ Cheese

▶ Yogurt

18

Milk

How many of
these dairy foods
do you eat?
Which one is your
favourite?

Ice cream

19

Cooking with milk

We use milk to make many other foods.

▶ *We use milk to make foods such as pancakes...*

▼ *cheese sauce...*

► *and milkshakes.*

Think about the
foods you eat.
Which ones are
made with milk?

I know that...

1 Most of us drink milk every day.

2 Milk helps our bodies in many ways.

3 Most milk comes from cows. They live on farms.

4 A cow makes milk when it has a calf.

5 Farmers milk the cows twice a day.

6 A tanker takes the milk to a dairy.

7 At the dairy, germs are killed so that the milk is safe to drink.

8 Milk is used to make butter, yogurt, cheese and ice cream.

9 We use milk to make other foods, such as pancakes, cheese sauce and milkshakes.

Index

About this book

I Know That! is designed to introduce children to the process of gathering information and using reference books, one of the key skills needed to begin more formal learning at school. For this reason, each book's structure reflects the information books children will use later in their learning career – with key information in the main text and additional facts and ideas in the captions. The panels give an opportunity for further activities, ideas or discussions. The contents page and index are helpful reference guides.

The language is carefully chosen to be accessible to children just beginning to read. Illustrations support the text but also give information in their own right; active consideration and discussion of images is another key referencing skill. The main aim of the series is to build confidence – showing children how much they already know and giving them the ability to gather new information for themselves. With this in mind, the *I know that...* section at the end of the book is a simple way for children to revisit what they already know as well as what they have learnt from reading the book.